[H.A.S.C. No. 114–132]

AIR DOMINANCE AND THE CRITICAL ROLE OF FIFTH GENERATION FIGHTERS

HEARING

BEFORE THE

SUBCOMMITTEE ON TACTICAL AIR AND LAND FORCES

OF THE

COMMITTEE ON ARMED SERVICES HOUSE OF REPRESENTATIVES

ONE HUNDRED FOURTEENTH CONGRESS

SECOND SESSION

HEARING HELD
JULY 13, 2016

U.S. GOVERNMENT PUBLISHING OFFICE

20–819   WASHINGTON : 2017

For sale by the Superintendent of Documents, U.S. Government Publishing Office
Internet: bookstore.gpo.gov   Phone: toll free (866) 512–1800; DC area (202) 512–1800
Fax: (202) 512–2104   Mail: Stop IDCC, Washington, DC 20402–0001

CONTENTS

AIR DOMINANCE AND THE CRITICAL ROLE OF FIFTH GENERATION FIGHTERS

HOUSE OF REPRESENTATIVES,
COMMITTEE ON ARMED SERVICES,
SUBCOMMITTEE ON TACTICAL AIR AND LAND FORCES,
Washington, DC, Wednesday, July 13, 2016.

The subcommittee met, pursuant to call, at 2:50 p.m., in room 2212, Rayburn House Office Building, Hon. Michael R. Turner (chairman of the subcommittee) presiding.

OPENING STATEMENT OF HON. MICHAEL R. TURNER, A REPRESENTATIVE FROM OHIO, CHAIRMAN, SUBCOMMITTEE ON TACTICAL AIR AND LAND FORCES

Mr. TURNER. The hearing will come to order. The subcommittee meets today to receive testimony on the need for air dominance and the critical role of fifth generation strike fighters. I want to welcome our distinguished witness for today, General Herbert J. "Hawk" Carlisle, Commander of Air Combat Command [ACC], United States Air Force. General Carlisle, we thank you for your service, and we look forward to hearing from you in your testimony today.

This hearing is the second of two important oversight hearings this subcommittee has held this year on the requirement of air dominance and the critical importance of fifth generation fighters in addressing current and emerging threats.

At our previous hearing held at the National Museum of the United States Air Force at Wright-Patterson Air Force Base on June 18, the witness, Major General Jerry Harris, the Vice Commander of the Air Combat Command, showed us a striking picture of one-half of an F–35 Joint Strike Fighter and one-half of a Chinese J–31 fighter juxtaposed jointly, appearing together. The similarities were shocking. It looked like one aircraft and left no doubt in anyone's mind that our adversaries are working very hard to challenge America's continued air dominance with fifth generation fighter programs of their own.

The last time the U.S. Air Force lost an aircraft in aerial combat was in 1972 when "DESOTO 03," an F–4E supporting Operation Linebacker II, was shot down by a North Vietnamese MiG–21. The advanced aircraft now under development by Russia and China signal their objective to end our 44-year advantage.

At the previous hearing, General Harris also noted that new surface-to-air missile systems now incorporate technologies allowing engagement at further ranges and in greater numbers. The sensitivity and accuracy of these new systems has also increased concerns regarding the unrivaled ability of our aircraft to access targets from anywhere, at any time.

Challenges to America's air dominance do not all originate from foreign shores. Some challenges are internal to the Air Force and the Department of Defense [DOD]. And one of the biggest challenges our Nation needs to overcome is the small size of today's Air Force.

For example, in 1991, during Operation Desert Storm, our Air Force had 134 fighter squadrons. Today, we are down to only 55 fighter squadrons. While the Department of Defense is no longer required to be able to defeat regional adversaries in large-scale campaigns on two fronts, we are losing our ability to do so on just one. We only produced 187 fifth generation F–22 aircraft. But that number was 194 aircraft short of the requirements for 381 F–22s. Unfortunately, the decision to stop F–22 production was a strategy driven by budgetary goals rather than one driven by the need to obtain a required capacity.

We don't want to make that same mistake with F–35 production that we made with our failure to produce enough F–22s. That is why the House-passed National Defense Authorization Act [NDAA] for Fiscal Year 2017 added 5 F–35As to meet last year's Air Force F–35A budget plan for 48 aircraft in fiscal year 2017, an unfunded requirement identified by the Air Force Chief of Staff. The House bill also added additional F–35Bs and Cs for the Navy and Marine Corps, also unfunded requirements identified by the Navy and Marine Corps.

April 15, 1953, is a significant date for the U.S. Air Force. It is the last time U.S. ground forces were killed as a result of enemy air attack when a North Korean P02 biplane strafed an Army tent on an island off the Korean peninsula. In the last 63 years, American air dominance has relentlessly safeguarded the lives of our Air Forces, provided freedom of maneuver and freedom from attack. I am confident we will do so now and in the future, but we must remain committed to providing the necessary resources to provide this capability, capacity, and readiness necessary to accomplish the critical mission of maintaining air dominance.

I would like to now recognize my good friend and colleague from Massachusetts, Ms. Niki Tsongas, for her opening comments.

[The prepared statement of Mr. Turner can be found in the Appendix on page 23.]

STATEMENT OF HON. NIKI TSONGAS, A REPRESENTATIVE FROM MASSACHUSETTS, SUBCOMMITTEE ON TACTICAL AIR AND LAND FORCES

Ms. TSONGAS. Thank you, Mr. Chairman. And good afternoon to you, General Carlisle. I am sorry we kept you waiting a little bit. But thank you so much for your service to our country and for being here to talk with us about this very important topic. Because what does bring us here today is the recognition that our Nation's Air Force faces a growing set of diverse and complex challenges around the world. To meet these threats and to maintain air dominance, the United States needs an Air Force with a range of capabilities to counter increasingly contested air environments and fighter advancements being developed by our adversaries. As we all know, the Air Force is in the midst of an ambitious modernization program driven, in part, by the age of many of its major aircraft fleets. Today, four major programs are in procurement, and five more are in research and development.

This is long overdue, as many airpower priorities have been deferred over the past decade in favor of ground force investments due to our engagements in Iraq and Afghanistan. Simultaneously, replacing such advanced programs is never easy. But this job is certainly made harder by the constraints placed on the Air Force by the Budget Control Act and a series of unpredictable budget deals over the past several years. In this resource-constrained environment, I look forward to hearing more today about how the Air Force prioritizes its major modernization programs, and how it aims to achieve a balanced set of capabilities to meet emerging threats. In addition, I hope to hear what Congress, we in Congress, might be able to do to help solve issues you are facing. I look forward to your testimony. Thank you for being here, and I yield back.

Mr. TURNER. Thank you. General Carlisle.

STATEMENT OF GEN HERBERT J. "HAWK" CARLISLE, USAF, COMMANDER, AIR COMBAT COMMAND, U.S. AIR FORCE

General CARLISLE. Chairman Turner, Ranking Members, and distinguished members of this subcommittee, it is a distinct pleasure to be here today with you to have this discussion. Thank you for the opportunity to discuss the importance of air superiority. As commander of Air Combat Command, I am responsible for many of the combat missions that our Air Force takes on. However, air superiority deserves special attention. It is the top stage setter for success on the battlefield; the mission that the Air Force must take on first and the Air Force mission that we must do right 100 percent of the time.

I am grateful the subcommittee shares an interest. And I am certain that our concern will advance the capabilities presented to combatant commanders. Future U.S. Air Force air superiority demands a full-spectrum force with capability beyond opponents, capacity to defeat emergent threats, and ready for worldwide deployment any time.

Currently, under BCA [Budget Control Act], I am unable to resource all three pillars of that mission: capability, capacity, and readiness. The Combat Air Force, to include air superiority, has been a bill payer in the last five budget cycles. For example, F–35 combat squadrons will be reduced from a planned 32 squadrons in 2028 to 16 in 2028.

The U.S. Air Force is the smallest, oldest, and busiest we have ever been. We are successful on the backs of our airmen. We prioritize the current fight at the expense of preparing for high-end operations. Today, our fighter force is less than 50 percent combat ready for full-spectrum operations. Our newly appointed Chief of Staff stated the most pressing challenge is the rise of peer competitors with military capabilities rivaling our own. Formerly accessible areas to the United States are now contested as Russian S–400 missiles and China's J–20 and J–31 aircraft are fielded. Our main air-to-air missile entered service in 1991. And our fifth generation aircraft still employ those fourth generation weapons, like the AMRAAM [Advanced Medium Range Air-to-Air Missile].

Our fifth generation fleet can only consist of F–22s until we reach sufficient numbers of F–35s to add to that fifth generation capability. Total F–35A requirement remains 1,763 aircraft, based on acquisition schedules on a projected fighter service life. An annual production of 60 F–35s right now strikes the right balance between cost and capability and the legacy aircraft aging out.

Looking further, we really must start now to devise our next generation answers and our next generation capability. We recently completed the Air Superiority 2030 Enterprise Capabilities Collaboration Team that concluded that there is no silver bullet solution, but, in fact, a multi-domain family of capabilities is ultimately the way we need to project and the way we need to win the fight in the future.

I look forward to an ongoing partnership with this subcommittee. And I thank all of you very much and the members of this entire committee for their dedication to air superiority the mission, our Armed Forces, and to our entire Nation. I welcome any questions from the committee chairman and everyone else on the committee. And I respectfully request my written testimony be entered into the record. Thank you very much for your time today. And it is an honor and a privilege to be here with you to answer any questions you might have. Thank you, Mr. Chairman.

[The prepared statement of General Carlisle can be found in the Appendix on page 25.]

Mr. TURNER. General, thank you. I have got a couple real quick questions that relate to issues that we are going to be facing in conference with this 2017 National Defense Authorization Act. And I would like your perspective to assist us in the negotiations over those provisions. One is that some have advocated that the F–35 Joint Program Office [JPO] be disestablished 180 days after the Milestone C decision in fiscal year 2019. Would Air Combat Command be in favor of devolving the JPO responsibilities to the services after Milestone C?

General CARLISLE. Sir, that is not something that we are in favor of. I truly believe that the JPO, the Joint Program Office for the F–35 has done a tremendous job. And they have done a very difficult job with three separate services as well as our partner nations, and FMS [foreign military sales] customers that are purchasing the aircraft. I do believe that the program office will evolve and change and there will be added responsibilities and added requirements on the individual services and their program offices.

So what I believe is the right answer for the future is an evolution of the Joint Program Office to one where some of their responsibilities and some of the things that are done are done by individual—either by the U.S. DOD, or by individual services. But the program office, I believe, still needs to be intact.

Mr. TURNER. Well, it certainly is a system that is currently working. And it does seem as if the risk would be too great to devolve that to the services.

General CARLISLE. Yes, sir.

Mr. TURNER. In addition, then some have advocated that the F–35 follow-on modernization program be treated as a separate major defense acquisition program, or MDAP. We had this issue in our markup, and it is not included—it was offered as an amendment. We realize that this generates from a GAO [Government Accountability Office] study, but the information that caused the House side to not adopt this policy decision was that it would cost about $13 million and delay Block 4 capabilities to the warfighters by 6 to 12 months. What are your thoughts and considerations concerning a major defense acquisition program for the modernization?

General CARLISLE. Mr. Chairman, I do not believe we should have a separate major defense acquisition program for the Block 4 update. It is incredibly critical that we get that capability. And delaying it only moves our capability to defeat potential threat to the future farther to the right. So with respect to time from an operator's standpoint, I truly believe we need to keep that as the program as part of the overall F–35 program, mostly because of the time. The fact is, and the JPO is better equipped to answer that, but it looks like it would add money and time. And both of those are things that we really can't afford to do.

Mr. TURNER. Money and time are both elements that this program has been criticized for. To take an administrative bureaucratic step that increases those certainly would impact the overall program. I appreciate your comment.

You know, General, actually my hometown newspaper in Ohio reported on a poll that had been done in Ohio, stating that 60 percent of the people were not in favor of the F–35. They were in favor, instead, of modernizing existing aircraft, pitting the F–35 against modernization programs for our current inventory. Obviously, in our prior hearings, it has been addressed that that is an impossibility, that one cannot modernize the existing aircraft in order to accomplish what is necessary with the capability of the F–35.

But clearly, we are still missing something, General, that the overall public in getting that phone call on the poll still does not understand that the leap that the F–35 is going to provide, and the risk of what our adversaries are developing and that the F–35 will face. How can you help us with that today?

General CARLISLE. Mr. Chairman, we have to be better at telling the story. The fact of the matter is that our adversaries have seen how successful we are. They have watched what has happened over the past 25 years. And they know if we dominate the airspace, that we can win any conflict and that we can be overwhelmingly lethal against adversaries. They know that and they are trying to counter that. And that is why you see things like you mentioned in your opening statement about the J–31, the J–20, the PAK FA from the Russians, the S–400, follow-on to those missiles. People know what we can do when we have air dominance. And they are trying everything in their power. And if we stay at our current technology, then we are just ceding that ground to our adversary, and we can't afford to do that. It would be—I mean, to take an analogy, if you took an old 1980s flip phone and tried to turn it into an iPhone 6, you can't do that. It is just not capable.

And if you look at the capabilities that the F–35 brings, it is centrifusion; it is low observable capability; it is a situational awareness for the pilot in the cockpit that is an order of magnitude different than its predecessors. And that is something you can't build onto or retrofit back into a previous generation airplane. You really need to go to the next generation.

Mr. TURNER. General, thank you. Turning to Niki Tsongas.

Ms. TSONGAS. Thank you, General. And to follow up on Mr. Turner's question and observation, I think we all appreciate, or have come to appreciate the difference that the Joint Strike Fighter will make. But conveying that to the average citizen, who really may not even understand what air dominance is, and how this new generation of airplane helps to achieve that, that is the challenge I think we all face here, especially in light of all the cost issues that have emerged, how long it has taken to develop this great capability. So, I think it is a tough one. And I think something that is worth considering how to better communicate the challenges we have and the difference it makes.

General CARLISLE. Yes, ma'am. I couldn't agree with you more. And I will tell you that part of it is us and the United States Air Force. We have to be better at telling the story. And I just got back from the Royal International Air Tattoo where we had both an F–35B and an F–35A. And it really does showcase the capabilities of the airplane and what it can do. We could talk about it with our partners and our FMS customers over there.

I think, you know, to some extent, frankly, we have been a little bit victims of our own success, because you see what has happened over the past 25 years, and we have had air dominance. But it has been from a lower capability threat. And we know that if you look at a resurgent and an increasingly aggressive Russia, you look at what is going on in the South China Sea and the East China Sea. If you look at what is going on with Iran and the weapon systems they are buying, all of those point to the fact that our adversaries know what we are capable of and they are doing everything in their power to counter it. We have to be better at telling that story, ma'am.

Ms. TSONGAS. I think that is true. I just wanted to return, though, to a slightly different question, and that is that of maintaining a diverse array of capabilities as you do seek to modernize. So recently, General, the Congressional Research Service indicated that four procurement programs accounted for 99 percent of the Air Force's aircraft acquisition budget in fiscal year 2016. And over the next several years, the Air Force plans on transitioning other important programs into procurement, including the JSTARS [Joint Surveillance Target Attack Radar System] recapitalization program as well as other modernization priorities. So my question is, General, I am curious about your thoughts about the Air Force's plans to invest more resources into bringing these programs online while continuing to acquire current programs that have proven susceptible to cost increases, as we know, higher than predicted operations and sustainment costs and other delays.

General CARLISLE. Yes, ma'am. Well, clearly, I think in the acquisition cycle and the procurement and setting requirements and holding requirements steady, we have to continue to get better. And in some of those cases, those cost increases were not necessarily a program problem, but as things changed—when program—acquisition programs get drawn out, then things change over time with respect to adversary technology and adversary capability and our own.

So, I think we really do have to get better at holding requirements steady and making the acquisition process more agile and flexible, so we can make—we can acquire programs kind of in the timeline we want to, on budget and on schedule, and make that schedule agile and flexible to make that happen.

Clearly, I think if you look at what has happened in the past 20 to 25 years, we kind of ended up in a position where we kind of stopped procuring in the 1990s. Frankly, we—in the peace dividend, the world changed drastically after—if you look at the air war over Serbia, our allied force combined with what happened on 9/11, that kind of changed the focus to a large extent, and we have concentrated pretty significantly with respect to investment on the current fight we are in.

At the same time, we see the adversary capability grow with the potential adversaries out there. So I believe that we have to prioritize. I think our nuclear enterprise has to be part of that. I think our space enterprise has to be part of that. I think our cyber enterprise has to be part of that. But we also have to modernize our capability to do the core function of the United States Air Force. And one of those is air dominance.

So modernization of F–22, procurement of F–35 weapons, and new generation weapons, and the right number of weapons to complete the task we need to. We have to prioritize those and then we have to figure out what the Nation wants us to do and what resources we need to be able to get to that point.

Ms. TSONGAS. Well, it is a daunting set of challenges in the face of very real fiscal constraints. And none of those things is inexpensive.

General CARLISLE. No, ma'am.

Ms. TSONGAS. So I don't envy you with your challenge. With that I yield back.

Mr. TURNER. Mr. Cook.

Mr. COOK. Thank you, Mr. Chairman. As chair of the dinosaur caucus, some of my questions might reflect that I have been on this planet for a while. I am a little concerned about the F–35. And, you know, I am all in on it and everything like that, but I stated, as a dinosaur, that, you know, supportive of the A–10. I really like the F–22, because of its proven track record. I know there has been some conversation about reopening that line again, particularly if the F–35 has more problems. And even, you know, the U–2 and things like this. I am not advocating bringing back the P–51 or the P–47 or the P–38. But, you know, the B–52 is still flying around. So conversation on the F–22, there has been some talk about that. And I think it will come up again if the F–35 has some prob- lems. And, as I said, in the dinosaur caucus, I support those air- craft that have a proven track record. So I really like it— you know, I am not a pilot. I don't know anything about it. You know, I can't even spell airplane. But can you comment on some of the conversations about the F–22 because——

General CARLISLE. Yes, sir.

Mr. COOK [continuing]. I liked it.

8

General CARLISLE. It is a fantastic airplane. It is absolutely the greatest air dominance airplane in the world today by an order of magnitude. Sir, I will tell you, first of all, the last person in the world that wants to get rid of any airplane in the United States inventory is Hawk Carlisle. I don't have enough capacity today. So, I love the fact that we are keeping airplanes around. And the fact of the matter is in the case of the A–10 that you talk about, it is a fantastic airplane. It does go a little bit to Congresswoman Tsongas' challenge, though, is how do you fit it all underneath the top line. You know, when we look at what we are challenged with, we are challenged with capability. So we have to get to that next generation capability, i.e., F–22s and F–35s, and eventually a B–21.

At the same time, we have to maintain capacity to meet all the demands around the world that are being asked from us. And I will tell you, with the world situation, those demands are going up. We average 10 fighter squadrons deployed at any given time 100 percent of the time in the United States Air Force, which is a huge commitment.

And then we have to maintain readiness. Our pilots and our maintainers and our airplanes have to be ready if something should happen be it anywhere in the world. So that is the balancing act we have to get to. The F–22 is a fantastic airplane. And as General Welsh said before he departed, I don't think it is a crazy idea to restart it. I do think that we probably would not bring an F–22 back in the form it is today. I think that is technology that is 30 years old, frankly. I think you may look at what the F–22 has and look at something—additive technology or things that you could potentially do different if you brought it back. The challenge with bringing it back, certainly in its current form, is the amount of time it would take to bring the subcontractors back on the line, get the tooling back up, start producing the airplane. What kind of cost that would be and how long it would be until you could get them. But I do believe that there is a potential, maybe, to look at what we have learned in the F–35 and what we have learned on the F–22. And maybe there is something in an F–22-like capability that we could bring that is the next generation, and the next capability and the next technology.

And the last thing I will say and then I—obviously I have a lot of opinions on this—is the F–35 is a fantastic airplane. It really is doing well. It is actually ahead of where the F–22 was in the same point in development that the F–35 is today and the F–22 was 10 or 11 years ago.

So I will tell you, I am very confident in that airplane. I am very close to declaring initial operational capability in that airplane because I believe in it. And the progress we are making, and the progress we have made even in the last year, is really tremendous. So I have confidence in it and I am very confident that we are not going to have additional problems in that airplane.

Mr. COOK. Well, thank you, General, for addressing that. The other thing, and I mentioned this before. You know, with the Canadians backing out of the F–35 buy, and who knows who would have predicted the U.K. [United Kingdom] and its political decision and everything like that. I am just a little nervous or worried about people that have committed as part of this buy, because it is going to influence the price and everything like that. And I am just hoping no one else decides to—who is going to pick up the slack and whether we could trust them or what have you with such an exceptional aircraft, if you could briefly comment on that.

General CARLISLE. Yes, sir. So, Congressman, I just spent the last week talking to many of our partners again. It was great out at the Royal International Air Tattoo at Fairford to have F–35Bs and F–35As flying there. We brought them over. We actually had the airplanes in country. They did fantastic. And, you know, I talked to the head of the Royal Canadian Air Force, and he thinks that decision is still open. He believes in the F–35, and from a military standpoint, he thinks his government is still potential. They are in a kind of competition now. Instead of a done deal, they are in a competition with some fourth gen [generation] airplanes.

And I will tell you the other thing, sir, that I spent a lot of my career in the Pacific. And if you look at Australia and Japan and Korea and Singapore, I think that market for that airplane is going up. I believe there is more and more enthusiasm, belief in it, and support for it. And talking to both the outgoing air chief from the Royal Air Force and the incoming air chief is they are going to buy 138, and my guess is they will buy more than that. And I think they will have a mix of both F–35Bs and F–35As at the end of the day.

Mr. COOK. Thank you. I yield back.

Mr. TURNER. Mr. Johnson.

Mr. JOHNSON. Thank you, General, for your service. And thank you for being here today. Is there anything critical about the F–22 to our tactical air superiority at this—as we proceed into the future? Is there anything critical about it?

General CARLISLE. Sir, there is. And that is, to continue the modernization. The airplane is a fantastic airplane, but as with everything, technology is evolving. We have a modernization program that includes some capability. We are continuously making our aircraft better. We are in a drop on the flight profile that is 3.2, that if you have seen what the F–22 has done in the Operation Inherent Resolve, it is just fantastic. We have been dropping SDBs, small diameter bombs, with great accuracy from that airplane. We have been able to penetrate airspace that other airplanes couldn't penetrate. So the criticality in the F–22 program today is to continue to modernize it, is to continue to add that capability as things go along, even the things that, like, low observable maintainability.

In the F–35, we developed this capability to rapidly take panels on and off and not have to do the whole low observable cure time in what is called a mighty boot, which is a capability to just put the parts back on. We are taking that from the F–35 and adapting that to the F–22, again, to continue those airplanes to maintain that, the greatest capability that we possibly can. So the one that I would really ask for is to maintain the modernization program. Mr. JOHNSON. And as we maintain that modernization program, would there be any need that you would see that we would need to take the training jets and convert them to combat capable aircraft?

General CARLISLE. Sir, I would love to do that. I would love to take the 43rd Fighter Squadron's jets down at Tyndall and upgrade them. They were very early model airplanes. And, again, with the production line shut down, we have to look at what the cost is to that. If it was a cost that was within reason and with everything else that I am trying to do I could do, I would consider doing that. We are, right now, looking very hard at what it would take to up-grade those airplanes to be in the most latest—latest block capa-bility. And we will look at that cost and we will come back, obvi-ously, to this committee and the Congress and talk about what the cost is and then what the benefit. But the more combat capable F–22s I have, the much happier I will be.

Mr. JOHNSON. Thank you. The F–35 procurement rates, are they sufficient to meet the requirement to reduce risks in potential com-bat with near-peer adversaries or in lower-risk combat environ-ments?

General CARLISLE. Sir, not yet. We need to get there. I believe that the number we need to get to is 60 a year. And I would like to do that as quickly as I can. We are not there yet. Some of the decisions to reduce the buy earlier were smart because we had—in the development of that program, we had a thing called con-currency where we were buying them and developing them at the same time. That is some of the early bad press that the F–35 got was we had a concurrency issue. So we slowed the buy rate to fix those problems, and we will go back and retrofit those early air-planes.

I believe now we are at the point where we can increase that buy rate, because those problems have been fixed and we don't see any coming in the future, and we have gotten through the concurrency part. So I believe that as soon as we can get to 60, the better.

Again, we are not there yet. We truly appreciate this committee and the House adding airplanes in the current budget. I am very much in favor of that. And, again, I would like to get to 60. Ulti-mately, I would like to get to 80 a year. But again, within all the priorities of the Air Force, we have to find out if—how we can fit that in. And there is so much going on in the recapitalization that that is the challenge that we will face.

Mr. JOHNSON. Thank you. As you mentioned in your testimony, the F–35's weapon system is a prime example of a weapon system with the ability to process large amounts and multiple sources of data. Clearly, these capabilities would create a significant advan-tage over our adversaries in the way that we track, target, and en-gage our enemies. Could you further elaborate on how the F–35 weapon systems work, and also, how do we ensure that the funding of this aircraft and its technology become a priority in the defense budget?

General CARLISLE. Yes, sir. The modern airplanes, like the F–35 and the F–22, and, in many cases, things like the next generation RPAs [remotely piloted aircraft] that we are going to develop over time, they are Hoovers for information. They just suck up tons of information because of the fused sensor suite and the amount of data they are able to collect and fuse. We need to get that informa-tion off board and take advantage of it for the entire capability from the tactical edge all the way back to the command and control and the decision makers.

In ACC, we are working very hard on what that networking and that off-board capability looks like in a thing called the combat cloud. And it is really about data to decision. So we take data. We use big data machine to machine, and we are able to use that in a security capability in a networked environment so that we can get that back to decision makers as well as to all the platforms in the tactical arena so everyone has the best information, and we can defeat our adversaries by knowing more sooner than they do and be able to react and force them into defensive mode. And we are very much doing that with the F–35. And, again, the F–35, F–22 both provide that capability. And we are working hard to make that happen.

With respect to the F–35, sir, we are—the place we are at in the program now, the Air Force is very close to initial operational capability or combat capability. And we are continuing to put that at the forefront of priorities to make sure we take advantage and we continue to develop that airplane and get to the Block 3F, which is ultimately the interim block we want to get to.

Mr. JOHNSON. Thank you, sir. And I yield back.

Mr. TURNER. Mr. Gibson.

Mr. GIBSON. Thanks, Mr. Chairman. I think—okay. Thanks. Ms. McSally. Did you want to go first?

Mr. TURNER. We had a joint dueling passing of time, and so, yes, I took Ms. McSally's signal to yield to Mr. Gibson.

Mr. GIBSON. The dinosaur caucus here. Oh, thank you very much. General, thanks for your service. And, you know, as the chairman mentioned, you know, we are in now the beginning process of the conference. So, you know, your testimony already is very timely and insightful. Thank you for that. Related, later this summer, I will be going out on a trip. And, you know, we are going to be focusing in on the European Reassurance Initiative. Mr. Cook will be going on it and some others. And so I am interested in hearing from you in the same vein or the same theme of telling the story, help us from your vantage point, explain what a CONOP [concept of operation] would look like that would be responding to—of course, we are in an unclassified setting, but help me ex-plaining this to my constituents on how the Air Force fifth genera-tion, how this is all put together as part of the joint team to be able to deter, and then if necessary conduct an operation. So I am inter-ested in hearing that from the—initially, the European perspective, but then any reinforcing that is brought to bear in the unclassified setting.

General CARLISLE. Yes, sir. And I would be more than happy at some point to come back and talk to the committee at a classified level as well, if that is something that the committee would so de-sire.

So the capability that those airplanes bring is the ability to pene-trate airspace. It is also a great messaging tool. Recently, we did what was called a Rapid Raptor where we deployed unannounced 12 F–22s to the European theater, and we moved them around the European theater, and they worked in coalition with fourth genera-tion aircraft, like A–10s and F–15s and F–16s. Hugely, hugely suc-cessful deployment. It is a deployment that demonstrates a capa-bility. A deployment that demonstrates resolve. And it is a deployment that demonstrates that we can put these airplanes where we want them when we want them there, in order to accomplish the mission.

Very much in the near future, after we are operational with the F–35s out of Hill Air Force Base, I would like to do that with the F–35s as well. One of the things that in Europe that I just talked to General Scaparrotti, the EUCOM [European Command] commander, and General Gorenc, the USAFE [U.S. Air Forces in Europe] commander, we would like to have F–35s, for example, do some Baltic air policing, where that is one of the missions that NATO [North Atlantic Treaty Organization] does up in the Baltic regions, in Estonia and Latvia and Lithuania, in that part of the world and put F–35s to demonstrate.

As I talked to the air chiefs over in Europe in this past week, all of them are very interested for their own countries to be able to see the visibility of that airplane out doing operational missions. Just like the F–22, getting the F–35 out there operationally conducting missions is very important. So—and the other part about the F–22 and the F–35, and the F–35 will be another example of this, it makes every other airplane on the battlefield that much better. It raises everybody's game because of the situational awareness it provides, the capabilities it can bring to bear, and the ability to change the fight with not—with a combination of fifth generation and fourth generation airplanes. So as we go forward, I think that is what we have to continue to demonstrate and then talk to people about it, what we can do with that, sir.

Mr. GIBSON. Thank you. And if I can just follow up on that comment, your conversations with the Supreme Allied Commander of Europe, and I am interested to know, in relation to some of the things you have done and what you are looking to do, the dynamic of how you think this is impacting—I mean, part of what we are trying to do obviously is reassure our allies, but we also want to see them bring more to the table here. We know we all need to step it up with regard to this. So in relation to the Air Force's piece of this and, you know, the increased experimentation in exercises, have you seen any change in the dynamic of the discussion of our friends and allies?

General CARLISLE. Sir, I have. And I will tell you, I will caveat it a little bit, because it is usually—I am usually talking to military members. And they love it when we come over. We put F–22s into Amari, Estonia, which was a tremendous, tremendous capability. We do the same with—we put A–10s in there. We put them into Romania, as well as on the—we flew with them in Germany. We just had a trilateral exercise at Langley Air Force Base where the French brought out Raphaels. The British, the RAF [Royal Air Force], brought out Typhoons and they flew with our F–22s. The more we do that, the more reassuring it will be for those nations, the more reassuring for their political as well their population. And I think if you—especially in Europe, and the same thing if you go to the Pacific, because what has happened in the South China Sea, and obviously the information that—the ruling that was just passed on by the International Court, all of those things, the tension that exists in the Pacific and in Europe with the things that are going on, the fact that we bring fifth generation capability, we

13

interact and they are interoperable with our friends and partners makes it—makes all those nations significantly more comfortable. And they truly, truly appreciate us being out there with this capability to interoperate with them.

Mr. GIBSON. I thank you for this, for enlightening us with regard to how we go about telling the story. And then I will just have my staff follow up with your staff before our team goes out on the CODEL [congressional delegation]. We will want at least like a laydown, maybe a brief. That wouldn't require you time——

General CARLISLE. Oh, no, sir. We would love to do that. Yeah, just let us know and we can talk about what we are doing, what we are doing in the future, and some things we are looking at. Be more than happy to do that.

Mr. GIBSON. Thank you, General, and thanks for your service. I yield back.

General CARLISLE. Thank you, sir.

Mr. TURNER. Ms. Graham.

Ms. GRAHAM. Thank you, Mr. Chairman. And thank you so much, General. I represent Tyndall Air Force Base.

General CARLISLE. Fantastic place. I spent a lot of time there.

Ms. GRAHAM. You did?

General CARLISLE. Yes, ma'am.

Ms. GRAHAM. It is a phenomenal place. And I had the opportunity to go up on a training mission with the F–22. I was in a T–38. Don't worry. I wasn't flying the F–22. And it is just an incredible, incredible airplane. I mean, I just can't even begin to describe what it can do in the air. It was amazing. And I am also very interested in the F–35. One of my earliest CODELs was to Eglin to learn about the F–35. Recently been down to Homestead Air Force Base in South Florida. They are potentially—I think they are in the running—I want to——

General CARLISLE. They are for one of the Guard units.

Ms. GRAHAM. Yes. Exactly. For F–35s. And if there is anything our office can do to help with information to encourage the placement of the F–35s there, of course, we stand ready to do that.

General CARLISLE. Yes, ma'am.

Ms. GRAHAM. That was a little plug, Mr. Chairman. He is not paying attention. But anyway, a question that your comments just brought to mind about the South China Sea, and I am actually getting ready to go on a CODEL to the RIMPAC [Rim of the Pacific] exercises.

General CARLISLE. Yes, ma'am. I participated in them many times. A fantastic exercise.

Ms. GRAHAM. I can't wait. I am very, very excited. But I am very concerned about—I am glad that our allies feel that we are working together well, and they are having their knowledge of what we are capable of doing and working together. But what are those that wish to do us harm, you know, what are their capabilities? China, Russia, Iran, what are we facing with their development of the technology? Because we know that they are watching us. And I would be curious to hear what you had to say about that.

General CARLISLE. Yes, ma'am. I could spend a lot of time, so I won't go into too level of detail. They have watched our success, they know how good we are, and they know that when we are going with air dominance then we pretty much can dictate the fight below us in a major contingency operation. They are doing everything in their power. And as the chairman mentioned, that picture of an F–35 and a J–31 where you have half of each, you can tell that they are copying us. You look at the PAK FA, which is the T–50, the Russian version of a stealth aircraft, you look at the missiles and what they are doing, and they are doing—all of our adversaries are doing two things. And that is where we come up with the term anti-access/area denial. They try to deny our ability to get into an area, try to keep us from being able to deploy there, and then once we get there, trying to restrict our ability to operate within that airspace.

F–22s and F–35s, in our modern systems what will eventually be B–21, the B–2, those are the answer to those challenges. And they are going to continue to modernize, they are going to continue to, as we have seen from our adversaries' cyber, they will steal technology so they avoid the challenges that we faced. And again, if you look at the J–31 and the F–35, it is not too hard to understand that they are successful at that.

So the answer, in my opinion, and what we are working on, is to continue to modernize, to continue to develop technologies, it is the third offset strategy that Secretary Carter talks about, and to continue to build on our capability. Because we as a Nation, and I truly believe this, I think many other nations, Russia and China in particular, copy very well. Original thought, they are not as good. And I believe that if you look at what our technology, what our industry does, what our airmen, sailors, soldiers, marines, and Coast Guardsmen can do, is take what we have and make it that much better. The greatest thing about watching F–22s and F–35s is handing them to captains and tech sergeants and seeing what they do with that capability. It exceeds anything we ever thought possible. And we are seeing that today in the F–35.

So I really believe our key to those adversaries that are continuing to try to deny us that is to continue to work that technology edge, experimentation, prototyping, systems engineering early to put technology into capability and then turn it over to our young men and women that are incredible when they get the opportunity to take advantage of what we give them and make it better than we ever thought possible. But our adversaries are there. You need only look at what the Russians are doing and what the PRC [People's Republic of China] is doing, and the fact that both those nations are selling that capability to many other nations that would wish us harm throughout the world. So it is incredibly important that we continue to stay on that edge, ma'am.

Ms. GRAHAM. Yes. And I really appreciate your comments. And I met with a great group from Eglin yesterday in my office, airmen. And I don't have time for this now, but if you could help follow up with our team about where we are with ALIS [Autonomic Logistics Information System] right now——

General CARLISLE. Yes, ma'am.

Ms. GRAHAM [continuing]. I would appreciate that.

General CARLISLE. Be more than happy. We are successful with ALIS on the backs of our airmen. Again, they are making it work, but we have to give them the right answer.

15

Ms. GRAHAM. I understand. Thank you very much. I yield back.

Mr. TURNER. Ms. McSally.

Ms. MCSALLY. Thank you, Mr. Chairman. General Carlisle, good to see you again.

General CARLISLE. Good to see you again, ma'am.

Ms. MCSALLY. As an airman, I am concerned, as you stated, that we are a victim of our own success, and that we have had air dominance—I mean, it has been like 60 years really since the last time an American was attacked from the air because we have been showing through the amazing capabilities of our airmen that we can protect them. Gaining and maintaining air dominance is a challenging task, especially as we see our adversaries, which you just mentioned, closing that gap, not just in capability, but also in numbers. And we need this fifth generation capability, but as you mentioned, we have got the smallest Air Force since its inception. And at some point, quantity has a quality all of its own. And so, we did recently get a letter from a Deputy Secretary of Defense re-lated to the numbers for F–35 staying with the total number, that 1,763 for the Air Force, saying it could go lower for budget reasons, or it could go up to keep pace with the threat. When I look at the threats, and we have had the briefings across the globe, and you have mentioned some of them, and we look at them in a classified level, I am deeply concerned about the numbers and our ability to be able to address varied simultaneous threats and have air domi-nance in all of them, all while our allies have dwindled in their budgets as well.

So, can you comment on just the number, and is there—is some-one looking at, are you looking at, and is there really a move to potentially increase the requirement in that number for the future?

General CARLISLE. Yes, ma'am. So I think that that is a great point. And the Air Force number is 1,763. And we believe that is the right number now. We are doing an analysis—and there is two different analyses going on. One was, initially the number was pre-dominantly driven by replacement for aircraft that we're aging out. That number of aircraft that have aged out has shrunk with the reduction in the size of our force. As the chairman said, we went from 134 squadrons, combat coded fighter squadrons in 1991, to 55 today.

So the number is smaller. But what we believe in Air Combat Command, and the Air Force believes, is that the number should be driven upon the threat and the environment that we are going to be asked to operate in. So I go to Congresswoman Tsongas's point, is we have to fit it all in there. But ultimately, this Nation and this body will determine what their military needs to be able to do, and then we need to have that capability in the systems to do that. I believe we are at the bottom edge of that. I think we need more capacity. If you look at what is happening in my force and how often I am rotating them through the Middle East, as well as the requirement I would have if something really bad happened with a near-peer competitor, be it a South China Sea environment, a Kaliningrad environment, or something to that effect, those two numbers, to me, is going to not only validate 1,763, it may be more and it may be what is next.

And I mentioned it just briefly about potentially whatever is next in our next gen capability, whether it is an F–22-like and we take that capability or whatever comes out of that, we have to have the capacity to meet the demands of the combatant commander to do what this Nation asks us to do.

Ms. McSALLY. I agree with you. And I realize you have constraints up the chain of command, but I think it would be helpful for us to know, based on the threat the number is this and then this is what we are saying we can afford, right, so at least we have an honest discussion and an understanding of numbers being driven by budget or driven by threats and capabilities. When it comes to—you mentioned we need a multi-domain family of capabilities, I think, and right now, that is fourth generation. You know, we have the F–22 with old technology. As we have had meetings with a number of the combatant commanders and others, again in classified settings, there has been a discussion—is there a way for us to kind of have a 4.5 or a 4.3, you know, or maybe taking some F–16 Block 50s and doing—I know you said you can't turn a flip phone into an iPhone. But is there anything we can be doing to augment the challenges we are having financially with the F–35 to, you know, create a 4.2 or 4.5 to augment so that we can have that multidimensional capability?

General CARLISLE. That is a great question. And ma'am, I guess my challenge is, I think modernization of fourth gen is important to continue to put that technology, just like it is for the F–22. I believe we need to continue to modernize fourth generation capability as well. The challenge that I would face is, if I bought new 4.5 generation aircraft, I don't know how I would do that and still buy fifth gen and what is next. You know, we are very close to getting the F–35 costs down into the $80- to $85 million, which is a very good cost for that airplane. And most fourth generation, or 4.5 generation airplane, would be in that same vicinity, in that same area. So I believe that the two things, from my perspective, that are most important is get the buy rate up on the F–35 so I buy them more quicker. That is as important as the end number is. I need to get to 60 if I can. And then we need to devise in a multi-domain is what is next? How do we continue to stay in front of our adversaries? How do we use space and cyber and surface and subsurface combined with air and the iCloud, the idea of the combat cloud technology to be able to defeat adversaries from different domains that they don't expect it to happen with the decision advantage inside of that.

Ms. McSALLY. Thank you. Mr. Chairman, can I ask one more quick question?

Mr. TURNER. Sure.

Ms. McSALLY. Okay. Thanks. I just want to go back to, I know this isn't a readiness hearing, but I think it is important just for a second for you to talk a little bit more about—you said less than 50 percent of our current fighter forces are ready to deploy for full range of combat missions, right? Less than 50 percent. Can you elaborate on that? You and I have talked about this a lot.

General CARLISLE. Yes, ma'am.

Ms. McSALLY. Some of that is FMC [fully mission capable] rates; some of it is related to parts in older airplanes; some of it is related to pilot shortages. I mean, this is significant. If we had to go tomorrow and we needed air dominance for any of these scenarios, this is our main factor. So could you just comment on that? It is important.

General CARLISLE. Yes, ma'am. So there is a few different areas that we can address in the readiness. But at the end of the day, if you look at full-spectrum operations against a high-end adversary, less than 50 percent of our fighter force is trained, capable, ready, and resourced with the parts, the munitions, the maintenance manpower, to be able to fight the high-end fight. And, you know, people have asked, we have had readiness budgets—quote-unquote "readiness budgets"—and people continually ask me, well, when are we going to be ready? When are we going to get back to that 80 to 90 percent readiness we need? And at the current state, we never will. We are treading water. We are not going backwards, but we are not making any progress.

It is all those things. Because the capacity is so small, we don't have enough time. Because we are turning 10 squadrons over every 6 months into the Middle East as well as doing TSPs in the Pacific and TSPs in Europe, theater security packages.

So time is a factor. We don't have enough people. We are trying to get that maintenance manpower back up so we can generate the sorties. And then the training. We have to keep the—we have to keep everybody trained. And if they are never home, or when they are home they don't have enough flying hours, you can't train them to that high-end fight. So right now we are treading water with respect to that.

Ms. MCSALLY. Thank you, General. Thank you, Mr. Chairman, for letting me indulge in that.

Mr. TURNER. General, I appreciate your great description of the need for the F–35, and obviously the problems we are facing. We just had today our conference committee on the budget. We appreciate your statements on the constraints that you are facing. And we are certainly trying to advocate for higher top-end numbers that can help address some of those constraints. With that, I want to thank you, General, for appearing before us. And we will be adjourned.

General CARLISLE. Thank you very much, Mr. Chairman.

[Whereupon, at 3:42 p.m., the subcommittee was adjourned.]

APPENDIX

JULY 13, 2016

PREPARED STATEMENTS SUBMITTED FOR THE RECORD

JULY 13, 2016

Statement of the Honorable Michael Turner
Chairman, Subcommittee on Tactical Air and Land Forces
Hearing on Air Dominance and the Critical Role of Fifth Generation Fighters
July 13, 2016

The hearing will come to order.

The subcommittee meets today to receive testimony on the need for Air Dominance and the critical role of fifth generation strike fighters.

I want to welcome our distinguished witness for today:

- **General Herbert J. "Hawk" Carlisle, Commander of Air Combat Command, United States Air Force**

General Carlisle we thank you for your service and look forward to hearing your important testimony today.

This hearing is the second of two important oversight hearings the subcommittee has held this year on the requirements for Air Dominance and the critical importance of fifth generation fighters in addressing current and emerging threats.

At our previous hearing held at the National Museum of the United States Air Force at Wright-Patterson Air Force Base on June 18th, the witness, Major General Jerry Harris, the Vice Commander of Air Combat Command, showed us a striking picture of one half of an F-35 Joint Strike Fighter and one half of a Chinese J-31 fighter joined together.

The similarities were shocking. It looked like one aircraft, and left no doubt in anyone's mind that our adversaries are working very hard to challenge America's continued Air Dominance with fifth generation fighter programs of their own.

The last time the U.S. Air Force lost an aircraft in aerial combat was in 1972 when DESOTO 03, an F-4E supporting Operation Linebacker II, was shot down by a North Vietnamese MiG-21. The advanced aircraft now under development by Russia and China signal their objective to end our 44-year advantage.

At the previous hearing, General Harris also noted that new surface-to-air missile systems now incorporate technologies allowing engagement at further ranges and in greater numbers. The sensitivity and accuracy of these new systems has also increased concerns regarding the unrivaled ability of our aircraft to access targets from anywhere and at any time.

Challenges to America's Air Dominance do not all originate from foreign shores. Some challenges are internal to the Air Force and the Department of Defense, and one of the biggest challenges our Nation needs to overcome is the small size of today's Air Force.

For example, in 1991 during Operation Desert Storm, our Air Force had 134 fighter squadrons. Today we're down to only 55 fighter squadrons. While the Department of Defense is no longer required to be able to defeat regional adversaries in large-scale campaigns on two fronts, we are losing our ability to do so on just one.

We only produced 187 fifth generation F-22 aircraft, but that number was 194 aircraft short of the requirement for 381 F-22s. Unfortunately, the decision to stop F-22 production was a strategy driven by budgeting goals rather than one driven by the need to obtain a required capability.

We don't want to make the same mistake with F-35 production that we made with our failure to produce enough F-22s. That's why the House-passed National Defense Authorization Act for Fiscal Year 2017 added five F-35As to meet last year's Air Force F-35A budget plan for 48 aircraft in fiscal year 2017, an unfunded requirement identified by the Air Force Chief of Staff. The House bill also added additional F-35Bs and Cs for the Navy and Marine Corps, also unfunded requirements identified by the Navy and Marine Corps.

April 15th, 1953, is a significant date for the U.S. Air Force. It is the last time U.S. ground forces were killed as a result of enemy air attack, when a North Korean PO-2 biplane strafed an Army tent on Chodo Island off the Korean peninsula. In the last 63 years, American Air Dominance has relentlessly safeguarded the lives of our Armed Forces, provided freedom of maneuver and freedom from attack.

I am confident we will do so now and into the future, but we must remain committed to providing the resources necessary to provide the capability, capacity and readiness necessary to accomplish the critical mission of maintaining Air Dominance.

Before we begin, I would like to turn to my good friend and colleague from Massachusetts, Ms. Niki Tsongas, for any comments she may want to make.

PRESENTATION TO THE
HOUSE ARMED SERVICES COMMITTEE
TACTICAL AIR AND LAND FORCES SUBCOMMITTEE
UNITED STATES HOUSE OF REPRESENTATIVES

SUBJECT: Future Air Dominance and the Critical Role of Fifth Generation Fighters

STATEMENT OF: General Herbert J. Carlisle, USAF
Commander, Air Combat Command

JULY 13, 2016

INTRODUCTION

Chairman Turner, Ranking Member Sanchez, and distinguished Members of the subcommittee, it is a distinct pleasure to be here with you this afternoon. Thank you for the opportunity to discuss the importance of Air Superiority and to highlight the jeopardy our nation will face if we do not continue a purposeful modernization of the Combat Air Force (CAF). As commander of Air Combat Command (ACC), the lead command for the CAF, I am responsible for organizing, training, and equipping the Air Superiority mission. Although it's one of several roles assigned to ACC, Air Superiority carries a special importance because it is the mission the Air Force must always address first. It is instrumental to achieving freedom of maneuver on the battlefield; not only in the air above, but also on the ground, on and under the sea, and in space. It is the precondition for success. It is the primary mission that must be accomplished to effectively impose our military will and might upon an active enemy.

Military leaders, myself included, along with this subcommittee, recognize the imperative for Air Superiority and its importance in ensuring our National Defense. American-led airpower has sustained the unprecedented advantage of Air Superiority in all conflicts since Vietnam. The thoughts that we are unable to destroy an enemy target at will from the air, or that our Armed Forces are at risk from enemy air attack have not crossed the minds of our leadership in decades. General Mark Milley, US Army Chief of Staff, recently stated, "[t]he fact of the matter is…when push comes to shove and bullets are actually flying and there are peoples' lives at stake…the United States Air Force has never failed me and it doesn't fail the Army." This enduring Air Force capability is provided by our airmen, their training and tactics, and our advanced aircraft, weapons, enabling systems, and battlefield networks.

However, nothing stands still in the nature of warfare and technology. Even today while we maintain Air Superiority in current areas of combat operations, we are flying near and within the weapons envelope of those that could test our dominance. The lead we have is shrinking as our near peer adversaries, and countries with which they proliferate, have developed, likely stolen, and fielded state-of-the-art systems. Highly advanced anti-access area denial (A2AD) technologies have made many formerly accessible areas contested. More and more inventories of aircraft, missiles, surface-to-air networks, and other weapons systems will soon challenge our ability to gain and maintain Air Superiority. America cannot effectively wield its military as an instrument of national power without the means to control the skies. When our means can be challenged, our ability to deter and dissuade washes away and is replaced with an adversary who sees a weakness; a weakness to be exploited and used to re-inspire thoughts of armed conflict.

As the threat will continue to evolve, so must we. Improvements and future investments are necessary to increase the capability, capacity and readiness of this indispensable mission. Limited resources, as always, will make this a challenge. Today's Air Superiority mission rests upon a mix of fourth and fifth generation fighters, supported by a highly refined command and control network, and flown by the world's best trained Airmen. However, balancing future capacity, capabilities, and readiness at the desired levels is near impossible within current financial constraints. ACC continues to design and advocate strategies to define requirements, increase acquisition agility, and reduce procurement timelines and life cycle costs, but these efforts can only carry so far. We are entering a period when the Joint force demands a full spectrum Combat Air Force, one that is modernized beyond the capabilities of our opponents, with the capacity to defeat emergent threats, and ready for short notice worldwide deployment. With the proper development, investment and commitment, the US Air Force can retain and even expand our Air Superiority capabilities, and bestow upon the next generation of Airmen, Soldiers, Sailors, and Marines what our predecessors bestowed upon us -- the freedom to maneuver our forces in the battlespace where we want, when we want.

LIMITED RESOURCES

Air Superiority plays an important role not only in prosecuting combat operations, but also in avoiding them. Although deterrence is usually thought of in reference to nuclear war, or the avoidance of, it also applies to conventional combat. The goal of Air Superiority is to be so capable that the enemy chooses not to fight. Limited resources have made this more challenging to accomplish. Undoubtedly, attaining the desired capacity, capability, and readiness levels will come at a cost. It is a traditional tradeoff between short term loss, or cost, and long term gain. The payoff is more than worth the effort. Overwhelming Air Superiority will give pause to any enemy bent on aggression.

The manner in which the Combat Air Force operates is based upon the three pillars of capability, capacity, and readiness. Currently, the CAF is unable to resource all three simultaneously. Over-focus on one exacerbates challenges for the other two. The CAF has been a bill payer for the last five budget cycles which in turn hinders capabilities. This includes six consecutive years of F-35 planning cuts that will reduce combat coded squadrons from the originally planned 32 to 16 by Fiscal Year 2028. In fact, since Operation Desert Storm the United States Air Force has divested 3,000 aircraft and 200,000 Airmen. This near 50 percent reduction brought combat coded fighter squadrons from 134 to 55, and the deployment rate has not changed placing that much more stress on our Force.

Weapons advancement is also key to assuring Air Superiority. For years, we have been able to conduct first look, first shot, first kill tactics. Our more powerful and sensitive radars tracked more targets further away and our missiles had a longer range and a higher kill probability. Today our primary air to air missile is the AIM-120, a medium range missile. Originally entering service on the F-15C in 1991, the AIM-120 today is increasingly challenged by adversary counter-measures and it limits our 5th Gen aircraft effectiveness. It also carries insufficient range versus newer long range adversary missiles and will soon require recapitalization. But the capabilities of our missiles are not the only limiting factor. Global precision attack weapons lost 24 billion dollars in funding over the last 5 planning cycles, which equates to 45 percent less weapons capacity. Our aircraft lose effectiveness when they run out of munitions due to lack of magazine depth and overall inventory. We are currently delivering 4th Gen weapons from 5th Gen platforms, and even those weapons inventories are being depleted beyond the current campaign requirements. We are experiencing this first hand with diminishing munitions in our current bombing campaign on the Islamic State. The former Air Force Chief of Staff, Gen Mark Welsh stated "We need funding in place to ensure we're prepared for the long fight. This is a critical need."

Today's Air Force is the smallest, oldest, and busiest we have ever been. Current operations tempo and fiscal limitations are a significant impact on full spectrum readiness, and are forcing us to reduce capacity while slowing the growth of new capabilities in order to meet budget constraints. We are compelled to prioritize training and funding for assigned counterinsurgency missions at the expense of our training for high-end operations. As a result, we cannot recover surge capacity for major contingencies and meet all the global demand with ready combat forces. The bottom line is that less than 50 percent of fighter squadrons are combat ready for full spectrum operations.

Sequestration has created an extremely difficult task to balance capacity, capability, and readiness. Air Force readiness remains at historic lows, partly driven by 25 years of continuous combat operations and also contributing factors by decisions made in response to sequestration. There are potential impacts to readiness that are worrisome, but the most detrimental impact of BCA level funding will be delays to CAF modernization efforts and purchases. This will further erode the already shrinking capability gap with our near-peer adversaries and in turn increase the risk of the Joint Force.

A time of fiscal constraint is the reality. We will continue to operate with reduced resources, but this is not the first time in our storied history that we have fought through financial hardships. We also saw declines in funding during periods of the Korean, Vietnam, and Cold War conflicts. It was a problem then and is one now. But, this is also why we are and always will be the greatest Air Force the world has known. Every time we face a problem, we find a solution and that is because of the ingenuity and innovation of our Airmen. This is truly our asymmetric advantage to keep us ahead of our adversaries. American Airmen are the best problem solvers in the world. We need to give them the resources to solve this one – the future of Air Superiority, and we need to do everything in our power to keep them in our Air Force.

CURRENT AIR SUPERIORITY THREATS AND CHALLENGES

During the Cold War, American Air Superiority had one mission and one clear adversary. In the years following the Cold War, the threat receded, Air Dominance became a buzzword, but our peers regrouped. After witnessing the lethal consequences sanctioned by American Air Superiority in multiple conflicts over the past 25 years, they grasped the reality that challenging our freedom of movement was key to slowing us down. It now comes as no surprise that our near peer adversaries' capabilities have been modernized to specifically counter and negate American capabilities. These new emergent threat systems range from 5th Gen fighters and integrated air defense systems to anti-space and standoff weapons. And whether these threats materialize in the Pacific or European AORs, they share very similar traits. Our next step in securing our ability to gain Air Superiority into the future is to counter these systems designed to counter ours.

Our shrinking capability advantage factors into adversaries' calculus – the smaller the gap, the less the deterrent. New threat surface to air systems now incorporate technologies allowing engagement at further ranges in greater numbers. The sensitivity and accuracy of these systems has also increased, calling into question the unrivalled ability of our aircraft to access anywhere at any time. Many now also claim the ability to acquire, track, and target low observable platforms like our stealth aircraft. Today's Surface to Air Missile threat is a combination of legacy, modernized legacy, and digital systems which can be linked to enhance cooperation and efficiency. But advanced technology is not the only way capability is increased. One can look no further than 1999 when a Serbian air defense battery operating a 1960s era SA-3 system with modern agile tactics, was able to adapt to the battlespace, and acquire then shoot down an F-117.

Although aircraft are some of the most expensive and challenging systems to develop and field, our competitors have made progress in the quest to match and counter American aerial capabilities. We are witnessing the emergence of advanced aircraft such as the T-50 from Russia and the J-20 and J-31 from China, with full expectations that foreign military sales are in their future. These new aircraft may possess levels of stealth, super-cruise, and advanced passive and active sensors that can pose problems to our dominance of the skies. They are also integrating innovative data-link technology similar to ours, which coupled with the internal carriage of newly developed long range active missiles, threaten the 44 year period since our last air to air loss.

Another strategically important issue threatening our ability to conduct Air Superiority is our access to and defense of deployed locations. Russian and Chinese ballistic and cruise missile modernization programs are far more robust and survivable today than they were ten or fifteen years ago. Ballistic missiles, armed with conventional warheads, are an incredibly capable platform that can negate our ability to use airfields and runways across the world, including the PACOM, EUCOM, and CENTCOM theaters. Advanced cruise missiles now include stealth technologies and increased range, and are a potent threat, especially considering they can be launched from a multitude of platforms and locations. It is important for the United States to be able to locate these threats and neutralize them left of launch. In order to make this possible, it will be imperative to develop future programs to enable a family of capabilities that includes a new counter-air aircraft and improvements in our C2, ISR, Space, and Cyber capabilities

Our near peers have also been very busy modernizing their capabilities to threaten our enabling technologies and systems such as the electromagnetic spectrum, space, and cyberspace. The main goal of these advanced programs is the denial of communications, and this includes the ability to kinetically or non-kinetically attack our space assets, jam and confuse our tactical and strategic communications networks, and hack our computer systems. Air Superiority, especially the American version, relies heavily upon fast and efficient communications. Attacks upon our enabling systems could reduce our Air Superiority assets from a totally integrated system to individual aircraft with radars. The American doctrine of centralized control, decentralized execution could be decapitated, and we would be left with systems stuck mimicking operational tactics from the Vietnam era, where visual acquisition and identification became the norm. Our competitors realize this area is one of our greatest advantages, and denying its use could be all the leverage they need to level the playing field.

PROCUREMENT AND DEVELOPMENT OF NEXT GEN SYSTEMS

During the 1950s our "First Offset Strategy" exploited our nuclear superiority to overcome the Soviet advantages. The "Second Offset Strategy" of the '70s and '80s focused on stealth development, precision guided munitions and the networks that enabled them. Our Next Generation systems are critical enablers for the "Third Offset Strategy". The Air Force will strengthen its military advantage by purposely employing this asymmetric strategy. Our next generation of systems will deliberately capitalize on our strengths and exploit adversary weaknesses. By fusing data to achieve automatic, near real-time information exchange for our combat forces, we will be able to more effectively employ current and next generation weapon systems. The F-35 is a prime example of a weapon system that is able to process large and multiple sources of data, analyze and then display it to the warfighter to create a decision advantage over the adversary. Having superior decision speed is how we position ourselves inside our adversary's decision cycle. A fully data- integrated fleet is a critical advancement and will maximize our ability to find, fix, track, target, and engage the enemy.

Currently, the US Air Force conducts the Air Superiority mission with a mix of 4th and 5th generation aircraft. Our 4th Gen fighter fleet consists of the F-15C, F-15E, and F-16. These aircraft play a significant role in the near term, especially in the capacity realm as we have few operational 5th Gen fighters. The role of our 4th Gen fighters will diminish over time due to two main reasons. The first is they will age out and be replaced by more capable F-35s. But more pressingly, our 4th Gen fighters are more increasingly unable to operate in highly contested environments where advanced air defense systems render them ineffective. The rate at which we procure F-35s will now have a significant impact on our Air Superiority capabilities as we cannot slow the rate at which the enemy develops and fields advanced area denial systems. Mitigating the effect of which our 4th Gen fighters age is possible, but it consumes scarce resources required to field 5th Gen aircraft and develop the next generation of capabilities required in the mid-2020s and beyond.

Our advanced 5th Gen fleet consists of the small number of F-22s. These multi-role aircraft out-class every adversary aircraft currently fielded, but must be modernized to keep pace with weapon systems that will be fielded in the near future. Eventually, as the threat evolves, even our 5th Gen fleet will not be able to operate in the high end of the operational environment. Additionally, the small number of F-22s we were able to procure leaves us with a less than ideal 5th Gen capacity until F-35s grow in sufficient numbers.

The F-35 acquisition schedule and projected service life of the remainder of the fighter fleet continue to drive a requirement for 1,763 F-35As to preserve capability and capacity over time. Currently, 48 F-35s are set to be produced annually, but to address our capacity and capability shortfalls, the desired production rate is 60. Delayed F-35 procurement forces the Air Force to extend legacy aircraft and accept increased readiness risk. We must find ways to reduce the time to field new capabilities. Procurement and development of next gen weapon systems are the best use of limited funds to ensure the defense of the nation.

CONCLUSION

The Chief of Staff of the Air Force commissioned the Air Superiority 2030 Enterprise Capability Collaboration Team (AS 2030 ECCT) to develop capability options to enable joint force Air Superiority in the highly contested environment of 2030 and beyond. One big takeaway from this study is that there is no "silver bullet". We will need to develop a family of air, space, and cyber capabilities to prevail in the highly contested Anti-Access/Area Denial environments of the future. Although a program is not yet in place, it will be paramount to continue modernizing our fleet, and progress to the next new counter-air aircraft that is more survivable, lethal, has a longer range, and bigger payload in order to maintain a gap with our adversaries. We will also need to continue to develop our C2, ISR, Space and Cyber capabilities. This multi-domain approach is resilient, enabling highly contested operations.

On December 27th, 1972, Major Carl Jefcoat and Lieutenant Jack Trimble launched in an F-4 Phantom call sign "DESOTO 03". It was Lt Trimble's 99th combat mission over North Vietnam, and would also be his last. Their Phantom had been escorting B-52s during Linebacker II. Attempting to engage an attacking MiG-21, Jefcoat and Trimble soon lost visual as the MiG disappeared into the clouds only to reappear on their tail. The MiG fired a missile that impacted their aircraft forcing them to eject. Jefcoat and Trimble became "guests of the north" as POWs in the infamous Hanoi Hilton and were repatriated several months later at the war's conclusion. 44 years ago, DESOTO 03 was the last United States Air Force air to air loss. Air Combat Command is tasked to ensure that never changes.

The United States Air Force has provided a dominant reign of American led Air Superiority for many years. The critical need for this capability will remain as our adversaries will continue to test the might of this great nation. The United States Air Force is positioned to supply what our country demands. Proper planning, resources, and investment will ensure freedom of maneuver for our forces well into the future.

I thank the committee for their service to the country, Armed Forces, and specifically to the importance of advancing Air Superiority mission set. I have no doubt that this collaboration will continue to propel our forces and the combat output needed to properly support the needs of our combatant commanders. I look forward to a continued partnership and the success it will bear for the Joint Force and the Nation.

General Herbert J. "Hawk" Carlisle

Gen. Herbert J. "Hawk" Carlisle Commander, Air Combat Command, Langley Air Force Base, Virginia. As the commander, he is responsible for organizing, training, equipping and maintaining combat-ready forces for rapid deployment and employment while ensuring strategic air defense forces are ready to meet the challenges of peacetime air sovereignty and wartime defense. The command operates more than 1,300 aircraft, 34 wings, 19 bases, and more than 70 operating locations worldwide with 94,000 active-duty and civilian personnel. When mobilized, the Air National Guard and Air Force Reserve contribute more than 700 aircraft and 49,000 people to ACC. As the Combat Air Forces lead agent, ACC develops strategy, doctrine, concepts, tactics, and procedures for air- and space-power employment. The command provides conventional and information warfare forces to all unified commands to ensure air, space and information superiority for warfighters and national decision-makers. The command can also be called upon to assist national agencies with intelligence, surveillance and crisis response capabilities.

Prior to assuming his current position, General Carlisle was the Commander of Pacific Air Forces; the Air Component Commander for U.S. Pacific Command; and Executive Director, Pacific Air Combat Operations Staff, Joint Base Pearl Harbor-Hickam, Hawaii. As the commander, he was responsible for Air Force activities spanning more than half the globe, leading approximately 45,000 Airmen serving principally in Japan, Korea, Hawaii, Alaska and Guam.

General Carlisle graduated from the U.S. Air Force Academy in 1978. He has served in various operational and staff assignments throughout the Air Force and commanded a fighter squadron, an operations group, two wings and a numbered air force. The general is a joint service officer and served as the Chief of Air Operations, U.S. Central Command Forward in Riyadh, Saudi Arabia. During that time he participated in Operation Restore Hope in Somalia. He also participated in Operation Provide Comfort in Turkey and Operation Noble Eagle. General Carlisle served on the Air Staff as Director, Operational Planning, Policy and Strategy, Deputy Chief of Staff for Air, Space and Information Operations, Plans and Requirements, and twice in the Plans and Programs Directorate. He also served as the Deputy Director, and later, Director of Legislative Liaison at the Office of the Secretary of the Air Force.

The general is a command pilot with more than 3,600 flying hours in the AT-38, YF-110, YF-113, T-38, F-15A/B/C/D, and C-17A.

EDUCATION
1978 Bachelor of Science degree in math, U.S. Air Force Academy, Colorado Springs, Colo.
1982 Squadron Officer School, Maxwell AFB, Ala.
1984 F-15 Fighter Weapons Instructor Course, Nellis AFB, Nev.
1988 Master's degree in business administration, Golden Gate University, San Francisco, Calif.
1991 Air Command and Staff College, Maxwell AFB, Ala.
1993 Armed Forces Staff College, Norfolk, Va.
1997 Army War College, Carlisle Barracks, Pa.
2002 National Security Management Course, Syracuse University, N.Y.
2005 Seminar XXI - International Relations, Massachusetts Institute of Technology, Cambridge
2007 Executive Course on National and International Security, George Washington University, Washington, D.C.

ASSIGNMENTS

1. May 1978 - November 1979, student, undergraduate pilot training, Williams AFB, Ariz.
2. November 1979 - January 1984, instructor pilot and flight examiner, 525th Tactical Fighter Squadron, Bitburg Air Base, West Germany
3. January 1984 - January 1986, Chief of Weapons and Tactics, 9th Tactical Fighter Squadron, Holloman AFB, N.M.
4. January 1986 - April 1988, Chief of Weapons and Tactics and flight commander, 4477th Test and Evaluation Squadron, Nellis AFB, Nev.
5. April 1988 - July 1990, Director, F-15 Multistage Improvement Program, Tactical Fighter Weapons Center, Nellis AFB, Nev.
6. July 1990 - June 1991, student, Air Command and Staff College, Maxwell AFB, Ala.
7. June 1991 - July 1993, Chief of Air Operations-Forward Element, Joint Operations Directorate, U.S. Central Command, Riyadh, Saudi Arabia
8. July 1993 - June 1995, operations officer, 19th Fighter Squadron, Elmendorf AFB, Alaska
9. June 1995 - July 1996, Commander, 54th Fighter Squadron, Elmendorf AFB, Alaska
10. July 1996 - June 1997, student, Army War College, Carlisle Barracks, Pa.
11. June 1997 - June 1998, Deputy Commander, 18th Operations Group, Kadena AB, Japan
12. June 1998 - March 2000, Commander, 1st Operations Group, Langley AFB, Va.
13. March 2000 - February 2001, Chief, Combat Forces Division, Directorate of Programs, Headquarters U.S. Air Force, Washington, D.C.
14. March 2001 - February 2003, Commander, 33rd Fighter Wing, Eglin AFB, Fla.
15. March 2003 - August 2004, Chief, Program Integration Division, Directorate of Programs, Headquarters U.S. Air Force, Washington, D.C.
16. September 2004 - April 2005, Deputy Director, Legislative Liaison, Office of the Secretary of the Air Force, Washington, D.C.
17. May 2005 - June 2007, Commander, 3rd Wing, Elmendorf AFB, Alaska
18. June 2007 - November 2007, Director, Operational Planning, Policy and Strategy, Deputy Chief of Staff for Air, Space and Information Operations, Plans and Requirements, Headquarters U.S. Air Force, Washington, D.C.
19. November 2007 - August 2009, Director, Legislative Liaison, Office of the Secretary of the Air Force, Headquarters U.S. Air Force, Washington, D.C.
20. September 2009 - December 2010, Commander, 13th Air Force, Hickam AFB, Hawaii
21. January 2011 - July 2012, Deputy Chief of Staff for Operations, Plans and Requirements, Headquarters U.S. Air Force, Washington, D.C.
22. August 2012 - October 2014, Commander, Pacific Air Forces; Air Component Commander for U.S. Pacific Command; and Executive Director, Pacific Air Combat Operations Staff, Joint Base Pearl Harbor-Hickam, Hawaii
23. October 2014 - present, Commander, Air Combat Command, Langley AFB, Va.

SUMMARY OF JOINT ASSIGNMENTS

June 1991 - June 1993, Chief of Air Operations-Forward Element, Joint Operations Directorate, U.S. Central Command, Riyadh, Saudi Arabia, as a major and lieutenant colonel

FLIGHT INFORMATION

Rating: command pilot
Flight hours: more than 3,600
Aircraft flown: AT-38, YF-110, YF-113, T-38, F-15A/B/C/D, and C-17A

MAJOR AWARDS AND DECORATIONS

Distinguished Service Medal with two oak leaf clusters
Legion of Merit with three oak leaf clusters
Defense Meritorious Service Medal
Meritorious Service Medal with three oak leaf clusters
Air Force Commendation Medal with oak leaf cluster
Joint Meritorious Unit Award with oak leaf cluster

EFFECTIVE DATES OF PROMOTION
Second Lieutenant May 31, 1978
First Lieutenant May 31, 1980
Captain May 31, 1982
Major March 1, 1989
Lieutenant Colonel June 1, 1993
Colonel Sept. 1, 1998
Brigadier General Aug. 1, 2005
Major General Dec. 10, 2007
Lieutenant General Sept. 2, 2009
General Aug. 2, 2012

(Current as of July 2015)

QUESTIONS SUBMITTED BY MEMBERS POST HEARING

JULY 13, 2016

QUESTIONS SUBMITTED BY MS. TSONGAS

Ms. TSONGAS. General, as you may be aware, I have been concerned about the increasing rate of physiological events being experienced by F/A–18 aviators in the Navy. While I am not aware of any similar incidents as of late in the Air Force, many of us on this subcommittee remember how the F–22 fleet was impacted by a similar rate of events several years ago. Are you aware of any similar trends amongst the Air Force fleet?

General CARLISLE. The Air Force is aware of the increased rate of physiological events experienced by F/A–18 aviators in the Navy. The Air Force is participating in collaborative efforts with the F/A–18 System Safety Working Group to determine root cause and corrective analysis. Since the F–22 Life Support System Task Force concluded its investigative effort and accomplished a requested congressional testimony, the Air Force has not experienced a rate of physiological events in any air-frame that is comparable to that of the Navy.

The Air Force did, however, see a relative increase in physiological events in the F–15C/D community that began shortly after a fatal crash in August 2014. Since fiscal year 2011, the community averaged approximately 5.3 hypoxic-like events per year. That rate increased to roughly 13 in FY 2015. Over the next 18 months F–15C/D pilots reported physiological events at an increased rate so, the Air Force chartered an Independent Review Team (IRT) to determine root cause/corrective analysis. Several factors with equipment and procedures unique to the F–15 were found and mitigation measures were identified by the IRT. F–15C/D event rates have normalized since May 2016 following implementation of mitigation proce- dures.. The findings of both the F–22 and F–15 investigations have been captured and shared with sister services where applicable, and the Air Force continues col- laborative efforts through system safety working group participation.

Ms. TSONGAS. General, one of the unique aspects of the F–35 is that it is essentially a very advanced sensor as well as being an advanced fighter. However, a sensor is only effective if it can talk to other platforms and pass the data it collects. What investments is the Air Force making in the other tactical fighters that will enable the force to maximize the capabilities of the F–35?

General CARLISLE. Link 16 is the designated primary tactical data link for exchange of information on the battlespace which the F–35 participates. The ACC/A3, in Nov 2014, validated an operational requirement for additional capabilities on all AF Link 16 platforms (Concurrent Multi-netting, Concurrent Contention Receive and Link 16 Enhanced Throughput). The increased network throughput provided by these enhancements supports the increased volume of information exchanges, such as the F–35 sensor information, to aid in prosecution of additional targets with greater success.

Ongoing terminal modernization and platform implementation paths require a holistic, "enterprise centric" approach. Common implementation of these capabilities in the Link 16 terminals and integrated on AF platforms, including the F–35, reduces the risk of losing a shared common tactical/operating picture, situational awareness, and desired mission effectiveness.

The Combat Cloud Operating Concept adds to the importance of the enterprise centric approach to future investments by describing the required capabilities need-ed to enable data sharing amongst the tactical edge platforms. The Combat Cloud concept for operations developed from a need for data sharing between 5th-to-4th and 4th-to-5th generation fighter and bomber platforms that minimizes AF and DoD duplication of effort. The current CSAF directed Agile Comms Capabilities Based Assessment is tasked to develop a solution to implement a Combat Cloud that can operate in contested airspace in 2025–2030, with primary focus on 5th gen plat-forms.

———

QUESTIONS SUBMITTED BY MR. YOUNG

Mr. YOUNG. One of the critical components of Fifth Generation fighters is the ability to fly in super cruise, or at above mach 1 speeds without using afterburners. Can you discuss the importance of this capability specifically, as well as the importance of using this capability in training? a. How many range complexes in the country offer Fifth Generation fighters the ability to super cruise?

General CARLISLE. The time domain can be a weapon at the tactical level. It is advantageous to be able to employ air power more quickly over a larger area than our adversaries. However, the advantages go beyond getting somewhere faster or going further in the same amount of time. Supersonic speeds are also advantageous both in terms of survivability and lethality. In a defensive situation, speed can make targeting more difficult— it confounds the solution required. Offensively, it provides additional kinematics to some of our weapons. When delivering air-to-ground ord-nance, additional speed allows further employment ranges and much desired stand-off from threats. In an air-to-air engagement, additional missile kinematics allows: longer employment ranges, earlier shot opportunities, and/or shorter time of flight (i.e. the missile impacts sooner). All highly desired in an air to air engagement.

The key to super cruise is the fact that the fighter does not require afterburners to fly supersonic. Afterburners are extremely inefficient. If a non-super cruise air- craft requires supersonic speed, it will use much more fuel achieving this airspeed and thereby reduce its range and/or time on station. Neither of which are desirable. However, a super cruise capable aircraft can retain it desired range and time on station while still reaping the tactical advantages of operating at supersonic speeds.

Our mantra is to "train the way you fight." The goal is to make the training as realistic as possible. The higher the correlation between training and combat, the better. This includes super cruise. The pace of an air to air fight is vastly quicker at supersonic speeds. If our airmen are not trained at this pace, they may "get be- hind the jet" and lose the fight. Our airmen must internalize this faster pace and be ready to execute instinctually. This can only be achieved through regular realistic training.

Super cruise can be utilized in any airspace that allows supersonic flight. In gen-eral, the majority of the supersonic airspace is our many offshore/overwater ranges. However, the ability to train using super cruise over land is much more limited. This capability is limited to large training complexes such as the Nevada Test and Training Range (NTTR), the Utah Test and Training Range (UTTR), and Alaska's Joint Pacific Alaska Range Complex (JPARC).

Mr. YOUNG. As you are well aware, Alaska is home to the Joint Pacific-Alaska Range Complex (JPARC). According to many, this is the best airspace for training in the world. Do you agree with this? a. Is there any other training airspace better suited for training the full capability of Fifth Generation fighters than the JPARC? b. Why is the JPARC particularly suited for Fifth Generation fighters and what can and should be done to improve the capability of the range?

General CARLISLE. The Joint Pacific-Alaska Range Complex (JPARC) is certainly one of the best training ranges in the world. There are several attributes a training range must possess to fully enable Fifth Generation fighter training.

One of the major capabilities of the F–22 is its' ability to super cruise or fly at supersonic speeds without the need for inefficient afterburner use. F–22s routinely execute tactics at supersonic speeds and rely on this capability to maximize our tac-tical advantage against our adversaries. Our forces must train realistically. This re-quires training at supersonic speeds. Outside the JPARC, the ability to employ at supersonic speeds is extremely limited. Currently the only airspaces available for supersonic training are limited to overwater/offshore ranges. This limits realistic air to surface employment. The Nevada and Utah Test and Training Ranges (NTTR and UTTR) are exceptions, but range time available is limited due to the high demand of supersonic overland airspace. It is also important to note that the areas within the NTTR and UTTR allowing supersonic employment is limited and does not en-compass the entire range.

When our Fifth Generation fighters employ at supersonic speeds, these speeds ne-cessitate additional airspace to fully realize the tactical advantages of supersonic employment—range measured not in miles, but hundreds of miles. The JPARC has over 62,000 square miles of airspace. This is over five times larger in area than the NTTR and the UTTR. The JPARC's large size enables the CAF's Fifth Generation pilots to hone their combat skills in the most realistic environment possible.

Unlike overwater/offshore ranges, the JPARC has several live air to surface em-ployment areas. The ability for CAF pilots to employ actual ordnance following a combat representative profile is imperative to training, maintaining, improving, and validating our wartime capabilities.

The overland nature of the JPARC also allows instrumentation. The JPARC has several portions that include instrumentation to provide time, space, and positional information for precise after action review. This ability to accurately review, assess, and debrief a mission is key to the USAF continuously updating and improving our tactics, techniques, and procedures (TTPs). In summary, the best training is the most realistic training. The JPARC enables the Fifth Generation forces to train at realistic speeds, employment ranges, to employ live ordnance, and is instrumented. This highly desired combination of attributes enables an extremely high correlation between training and combat. Thereby providing an environment for some of the best training in the world.

As stated above, the JPARC is undoubtedly one of the best training ranges in world. If ones defines improving the "capability of the range" as improving both the capacity and realism of the range, then investment in Live, Virtual, and Constructive (LVC) infrastructure will provide the most impact. Live training is actual pilots flying actual aircraft. Virtual training is actual pilots flying virtual aircraft—simulators. Constructive training is utilizing computer generated entities controlled by either a man-in-the-loop or autonomously by the computer. The lines that previously separated these training domains are beginning to blur. Technological advancements now allow airmen flying actual aircraft to train with airmen in simulators to a limited degree anywhere in the world, augmented by computer generated threats. This emerging capability leverages our Distributed Mission Operations (DMO), which links dislocated simulator sites together for mutual training, and our advanced datalink capabilities. By combining these capabilities/domains we can exponentially increase capacity by increasing participation via additional assets linked into the training scenario. An example would be live F–22s and F–35s flying on the JPARC training with an AWACS crew operating out of a simulator at Tinker AFB in Oklahoma. Adding constructive threats also increases realism. We can now "inject" constructive Surface to Air Missile sites (SAM), or threat aircraft to augment the training scenario's realism. An example would be adding constructive MiGs and advanced SAMs to the training scenario and datalinking to the live aircraft.

Mr. YOUNG. Can you please describe the importance of RED FLAG-Alaska, and how it is different but complimentary to RED FLAG-Nellis?

General CARLISLE. Both exercises provide similar world-class air combat training for our Combat Air, Space and Cyber forces, sister services, and allied partners from over 30 countries. While similar, each has certain advantages that benefit USAF Air, Space and Cyber forces. In general, RF–N offers more high-end, multi-domain integration with Air Operations Center (AOC) support, while RF–A offers a range whose dimensions better satisfy 5th generation fighter requirements. Although both exercises incorporate Live, Virtual, and Constructive (LVC) elements to enhance training, more work is necessary to unlock its full potential. Current efforts to achieve a more logical and supportable "strategic calendar" may allow scheduling our forces to take advantage of what both exercises offer. Second order effects of this include better mutual support and deconfliction between the training squadrons that run the events, better/more flexible support from enabling units, and less travel burden on the operational units themselves.

Mr. YOUNG. Alaska will soon become the premier location for combat-coded Fifth Generation fighters, with F–22s currently at Joint Base Elmendorf-Richardson and F–35s expected to arrive to Eielson Air Force Base in a few years. Can you talk about the importance of these aircraft training together, along with the 18th Aggressor Squadron at Eielson providing red air?

General CARLISLE. The combination of F–22s at Joint Base Elmendorf-Richardson, F–35s at Eielson Air Force Base, the 18th Aggressor Squadron also located Eielson, and access to the Joint Pacific Alaska Range Complex (JPARC) sets the stage for some of the best training to be found anywhere. This is important because as the USAF modernizes our fleet with 5th generation capabilities our tactics, techniques, and procedures (TTPs) need to modernize as well. The ability for F–22s and F35s to easily train together will enable our combat forces to quickly develop and continuously fine-tune these TTPs. In Alaska we will be able to train locally on a daily basis without requiring travel to get this valuable training.

Additionally, the collocated professional aggressor forces and access to the JPARC, one of the finest training ranges in the world, provides an optimum environment accelerating the development of these TTPs. The regular, repeatable, mixed-force training of our 5th generation forces in superb airspace, fighting the most capable aggressors will rapidly facilitate lessons learned that the entire CAF will apply and enable the USAF to remain the most capable and lethal force possible.

Mr. YOUNG. Further, given the large number of Fifth Generation fighters that will be based in Alaska—and a large amount of strategic airlift and Army airpower—does it not make sense to consider Alaska for the basing of the KC–46 tanker, especially given Alaska's 24/7 NORAD Alert Mission? a. What's the timeline and what would help Alaska's candidacy?

General CARLISLE. The Strategic Basing process for the beddown of the KC–46 tanker is run by Air Mobility Command (AMC). We have forwarded AMC this request.

Mr. YOUNG. From an ACC perspective, can you comment on the potential negative impacts of not fully funding the weather shelter for the second squadron of F–35s in Alaska? a. Based on your extensive experience in Alaska, is it sensible to leave $270 million ($100M/F–35) out on the runway during the winter time in Fairbanks, where temperatures can reach ¥40 to ¥50 for weeks at a time? b. What effects do you think these extreme cold temperatures would have on the F–35s?

General CARLISLE. It is not prudent to leave the F–35 out on the ramp in extreme temperatures and doing so could have a significant operational impact if not protected. If the second squadron at Fairbanks cannot shelter their aircraft during the harsh winter months, there is the potential that the squadron would not be able to deploy/employ in the timelines expected by PACOM.

Possible effects of cold soaking the aircraft include broken lines/seals, shorter times to perform maintenance tasks such as repairs, modifications, and inspections. In addition, the individual pilot and maintainer also have a limited exposure times due to the extreme elements that Alaska climatology presents.

Mr. YOUNG. It is clear that Russia is seeking to reassert its strength around the world, and specifically in the Arctic Region. Based on this resurgence, as well as an unpredictable North Korea and belligerent China, can you discuss the importance of positioning Fifth Generation fighters in the Asia-Pacific and Arctic Regions, and specifically in Alaska?

General CARLISLE. Positioning Fifth Generation fighters in the Asia-Pacific and Arctic Regions is of vital importance. Fifth Generation fighters will contribute directly to our Nation's defense in Alaska. Currently F–22s sit alert for Operation Noble Eagle and basing of other Fifth Generation assets will only increase our capability and capacity to deter Russian Long Range Aviation. Similarly in the Asia-Pacific region these Fifth Generation fighters will act as a deterrence to both North Korea and China while protecting America's partner nations as well as strategic interests in the region.

Mr. YOUNG. Given these grave threats in the Asia-Pacific region, what will happen if the Department of Defense continues to delay its acquisition and modernization of fighter aircraft, and specifically Fifth Generation fighters?

General CARLISLE. Continuing to delay the acquisition and modernization of fighter aircraft, specifically Fifth Generation fighters will have serious consequences if conflict erupts in the Asia-Pacific region. History has demonstrated that air superiority ensures victory. The U.S. military and coalition allies will face an anti-access, area denial (A2AD) environment that will require a fifth generation fighter to achieve air superiority and enable us to hold any ground target at risk at a time or place of our choosing with precision and persistence. Fifth generation fighters offer first look, first shot, first kill through stealth, maneuverability, multi-role capabilities in addition to fused sensors and avionics. They also bring decision and reaction dominance, flexibility and survivability over our adversaries.

Too small of a Fifth Gen fleet leaves the U.S. and our allies vulnerable to enemy attack. Additionally, too small of a Fifth Gen fleet eliminates the possibility of deterring an enemy bent on aggression. If we cannot establish Air Superiority, we cannot be successful in any of our missions, and our country and the free world will be placed in jeopardy.

Due to the limited numbers of F–22s, modernization of capabilities remain crucial to ensure the Combat Air Forces continue to dominate adversary weapon systems that will be fielded in the near future.

For six consecutive years, the F–35 has experienced cuts in planned procurement, resulting in reductions in combat coded squadrons from 32 to 16 by Fiscal Year 2028. Delayed F–35 procurement forces the Air Force to extend legacy aircraft and accept increased readiness risk. We must find ways to reduce the time to field new capabilities.

Equally important are advanced weapons for air-to-air and air-to-ground combat employment. We must pursue and field cutting-edge weapons to realize full combat capabilities of Fifth Gen platforms.

Delaying the modernization or acquisition of Fifth Gen fighters in the numbers necessary will increase operational risk to our forces and prevent our ability to achieve air superiority and provide global position attack capabilities to support our joint force, allies, and national interests.